Anonymous

An Irish Agricultural Delegate on the Agricultural Resources
of Canada

The report of Mr. Jerome J. Guiry, of Peppardstown, Fethard, Clonmel,

Ireland, on his visit to Canada in 1893

Anonymous

An Irish Agricultural Delegate on the Agricultural Resources of Canada
The report of Mr. Jerome J. Guiry, of Peppardstown, Fethard, Clonmel, Ireland, on his visit to Canada in 1893

ISBN/EAN: 9783337328665

Printed in Europe, USA, Canada, Australia, Japan

Cover: Foto ©berggeist007 / pixelio.de

More available books at **www.hansebooks.com**

AN IRISH AGRICULTURAL DELEGATE

ON THE

AGRICULTURAL RESOURCES OF CANADA.

THE REPORT

OF

Mr. JEROME J. GUIRY, of Peppardstown, Fethard, Clonmel, Ireland,

ON

HIS VISIT TO CANADA IN 1893.

Published by Authority of the Government of Canada (Department of the Interior).

FEBRUARY, 1894.

PART VI.

CONTENTS.

		PAGE
PREFACE		iii
MR. JEROME J. GUIRY'S REPORT		1
GENERAL INFORMATION ABOUT CANADA (Appendix A)		27
THE CANADIAN EXHIBITS AT CHICAGO (Appendix B)		35
MAP		to face 38

LIST OF ILLUSTRATIONS.

CANADIAN PACIFIC RAILWAY HOTEL, QUEBEC ...	1
EXPERIMENTAL FARM, OTTAWA	3
WINNIPEG	4
FARM SCENE, MANITOBA	6
WHEAT STACKS, MANITOBA	7
GRAIN ELEVATOR, BRANDON	8
CAMERON'S FARM, QU'APPELLE...	10
HYDE FARM, QU'APPELLE	11
CALGARY	15
RANCH SCENE, ALBERTA	16
AN ONTARIO FARM	20
A FARM-HOUSE, SOUTHERN MANITOBA	26

PREFACE.

In July, 1893, the High Commissioner for Canada, by direction of the Minister of the Interior, invited the following gentlemen (who are all connected with the agricultural industry in the different parts of the United Kingdom from which they come) to visit the Dominion of Canada, and report upon its agricultural resources, and the advantages the country offers for the settlement of farmers and farm labourers, and the other classes for which there is a great demand :—

Mr. A. J. Davies, Upper Hollings, Pensax, Tenbury, Worcestershire ; Mr. W. H. Dempster, Millbrook Lodge, Clarbeston Road, South Wales ; Mr. Alexander Fraser, Balloch, Culloden, Inverness, Scotland ; Mr. R. H. Faulks, Langham, Oakham, Rutland ; Mr. J. T. Franklin, Handley, near Towcester, Northamptonshire ; Mr. J. J. Guiry, Peppardstown, Fethard, Clonmel, Ireland ; Mr. Tom Pitt, Oburnford, Cullompton, Devon ; Mr. John Roberts, Plas Heaton Farm, Trefnant, North Wales ; Mr. Reuben Shelton, Grange Farm, Ruddington, Nottinghamshire ; Mr. Joseph Smith, 2, Mowbray Terrace, Sowerby, Thirsk, Yorkshire ; Mr. John Steven, Purroch Farm, Hurlford, Ayrshire, Scotland ; Mr. Booth Waddington, Bolehill Farm, Wingerworth, Chesterfield ; and Mr. William Weeks, Cleverton Farm, Chippenham, Wiltshire.

In addition, two other farmers—Mr. John Cook, of Birch Hill, Neen Sollars, Cleobury Mortimer, Shropshire ; and Mr. C. E. Wright, of Brinkhill, near Spilsby, Lincolnshire—visited the Dominion, under their own auspices, during 1893 ; and they have been good enough to prepare Reports of their impressions.

The Reports, if published together, would make a bulky volume. It has therefore been decided to divide them into the following parts :—

Part 1—The Reports of Messrs. Shelton, Waddington, Cook, and Smith.
Part 2—The Reports of Messrs. Franklin, Faulks, and Wright.
Part 3—The Reports of Messrs. Weeks, Pitt, and Davies.
Part 4—The Reports of Messrs. Roberts and Dempster.
Part 5—The Reports of Messrs. Steven and Fraser.
Part 6—The Report of Mr. Guiry.

Part 1 will be circulated in the following counties :—Northumberland, Cumberland, Durham, Westmoreland, York, Lancashire, Shropshire, Cheshire, Staffordshire, Derby, and Nottingham.

Part 2, in Lincoln, Rutland, Leicester, Northampton, Huntingdon, Cambridge, Norfolk, Suffolk, Essex, Hertford, Bedford, Bucks, Oxford, Berks, Middlesex, Surrey, Kent, and Sussex.

Part 3, in Warwick, Worcester, Hereford, Gloucester, Wiltshire, Hampshire, Dorset, Somerset, Devon, and Cornwall.

Part 4, in Wales ; Part 5, in Scotland ; and Part 6, in Ireland.

From whom Pamphlets obtainable. Any or all of these pamphlets, as well as other illustrated pamphlets issued by the Government, may be obtained, post free, by persons desiring to peruse them, on application to the Hon. Sir Charles Tupper, Bart., G.C.M.G., C.B., High Commissioner for Canada, 17, Victoria Street, London, S.W. ; to Mr. J. G. Colmer, C.M.G., Secretary, at the same address; or to any of the agents of the Canadian Government in the United Kingdom, whose names and addresses are as follows:—Mr. John Dyke, 15, Water Street, Liverpool; Mr. Thomas Grahame, 40, St. Enoch Square, Glasgow; Mr. E. J. Wood, 79, Hagley Road, Birmingham; Mr. P. Fleming, 44, High Street, Dundee; Mr. W. G. Stuart, Nethy Bridge, Inverness; and Mr. G. Leary, William Street, Kilkenny. Copies may also be obtained from the steamship agents, who are to be found in every village.

Land Regulations in various Provinces.   As the land regulations of the different Provinces are frequently referred to in the Reports, they are quoted in detail in the following paragraphs; but they are, of course, subject to alteration from time to time :—

Prince Edward Island.—The available uncultivated and vacant Government land is estimated at about 45,000 acres. These consist of forest lands of medium quality, the very best having, of course, been taken up by the tenants in the first instance, and their price averages about one dollar per acre. Parties desiring to settle upon them are allowed ten years to pay for their holdings, the purchase-money to bear interest at 5 per cent. and to be payable in ten annual instalments.

Nova Scotia.—There are now in Nova Scotia about two millions of acres of ungranted Government lands, a considerable quantity of which is barren and almost totally unfit for cultivation; but there is some land in blocks of from 200 to 500 acres of really valuable land, and some of it the best in the province, and quite accessible, being very near present settlements. The price of Crown lands is $40 (£8 sterling) per 100 acres.

New Brunswick.—Crown lands, of which there are some 7,000,000 acres still ungranted, may be acquired as follows :—(1.) Free grants of 100 acres, by settlers over 18 years of age, on the condition of improving the land to the extent of £4 in three months; building a house 16 ft. by 20 ft., and cultivating two acres within one year; and continuous residence and cultivation of 10 acres within three years. (2.) One hundred acres are given to any settler over 18 years of age who pays £4 in cash, or does work on the public roads. &c., equal to £2 per annum for three years. Within two years a house 16 ft. by 20 ft. must be built, and two acres of land cleared. Continuous residence for three years from date of entry, and ten acres cultivated in that time, is also required. (3.) Single applications may be made for not more than 200 acres of Crown lands without conditions of settlement. These are put up to public auction at an upset price of 4s. 2d. per acre; purchase-money to be paid at once; cost of survey to be paid by purchaser.

Quebec.—About 6,000,000 acres of Crown lands have been surveyed for sale. They are to be purchased from the Government, and are paid for in the following manner:—One-fifth of the purchase-money is required to be paid the day of the sale, and the remainder in four equal yearly instalments, bearing interest at 6 per cent. The prices at which the lands are sold are merely nominal, ranging from 20 cents to 60 cents per acre (15d. to 2s. 5½d. stg.). The purchaser is required to take possession of the land sold within six months of the date of the sale, and to occupy it within two years. He must clear, in the course of ten years, ten acres for every hundred held by him, and erect a habitable house of the dimensions of at least 16 ft. by 20 ft. The letters patent are issued free of charge. The parts of the Province of Quebec now inviting colonisation are the Lake St. John District; the valleys of the Saguenay, St. Maurice, and the Ottawa Rivers; the Eastern Townships; the Lower St. Lawrence; and Gaspé

Ontario.—Any head of a family, whether male or female, having children under 18 years of age, can obtain a grant of 200 acres; and a single man over 18 years of age, or a married man having no children under 18 residing with him, can obtain a grant of 100 acres. This land is mostly covered with forest, and is situate in the northern and north-western parts of the province. Such a person may also purchase an additional 100 acres at 50 cents per acre, cash. The settlement duties are—To have 15 acres on each grant cleared and under crop at the end of the first five years, of which at least two acres are to be cleared annually ; to build a habitable house, at least 16 feet by 20 feet in size ; and to reside on the land at least six months in each year. *In the Rainy River* district, to the west of Lake Superior, consisting of well-watered uncleared land, free grants are made of 160 acres to a head of a family having children under 18 years of age residing with him (or her); and 120 acres to a single man over 18, or to a married man not having children under 18 residing with him; each person obtaining a free grant to have the privilege of purchasing 80 acres additional, at the rate of one dollar per acre, payable in four annual instalments.

Manitoba and North-West Territories.—Free grants of one quarter-section (160 acres) of surveyed agricultural land, not previously entered, may be obtained by any person who is the sole head of a family, or by any male who has attained the age of 18 years, on application to the local agent of Dominion lands, and on payment of an office fee of $10. The grant of the patent is subject to the following conditions having been complied with :—

By making entry and within six months thereafter erecting a habitable house and commencing actual residence upon the land, and continuing to reside upon it for at least six months in each year for the three next succeeding years, and doing reasonable cultivation duties during that period.

Persons making entry for homesteads on or after September 1st in any year are allowed until June 1st following to perfect their entries by going into actual residence. The only charge for a homestead of 160 acres is the entrance fee of $10. In certain cases forfeited pre-emptions and cancelled homesteads are available for entry, but slightly additional fees, and value of improvements thereon, if any, are demanded from the homesteader in each case, and when abandoned pre-emptions are taken up they are required to perform specified conditions of settlement. Full information can be obtained from the local agents. In connection with his homestead entry the settler may also purchase, subject to the approval of the Minister of the Interior, the quarter-section of the same section, if available, adjoining his homestead, at the Government price, which is at present $3 per acre. In the event of a homesteader desiring to secure his patent within a shorter period than the three years, he will be permitted to purchase his homestead at the Government price at the time, on furnishing proof that he has resided on the land for at least 12 months subsequent to date of entry, and has cultivated 30 acres thereof.

The following diagram shows the manner in which the country is surveyed. It represents a township—that is, a tract of land six miles square, containing 36 sections of one mile square each. These sections are subdivided into quarter-sections of 160 acres each, more or less.

TOWNSHIP DIAGRAM.

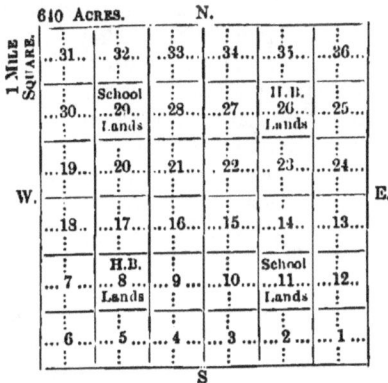

			N.			
	...31..	.. 32..	..33...	...34...	...35...	...36...
	...30...	School 29.. Lands	...28...	...27...	H.B. 26... Lands	...25...
	...19...	...20...	...21...	. 22...	.. 23...	...24...
W.	...18..	...17...	...16...	...15...	...14..	...13...
	... 7 ...	H.B. 8 .... Lands	...9 ...	...10...	School 11... Lands	...12..
	... 6 ...	...5 ...	...4 ...	...3 ...	...2 ...	...1 ...
			S			

610 ACRES. — 1 MILE SQUARE. — E.

The right of pre-emption has ceased to exist, having been altogether discontinued since 1st January, 1890.

Information respecting timber, mineral, coal, grazing and hay lands, may be obtained from any of the land agents. Homesteaders in the first year of settlement are entitled to free permits to cut a specified quantity of timber for their own use only, upon payment of an office fee of 25 cents.

It must be distinctly understood that the land regulations are subject to variation from time to time. Settlers should take care to obtain from the land agent, when making their entry, an explanation of the actual regulations in force at that time, and the clause of the Act under which the entry is made endorsed upon the receipt, so that no question or difficulty may then or thereafter arise.

British Columbia.—In this province any British subject who is the head of a family, a widow, or a single man over 18 years, or an alien proposing to become a British subject, may acquire the right from the Provincial Government to pre-empt not more than 160 acres of Crown lands west of the Cascade Range, and 320 acres in the east of the province. The price is 4s. 2d. an acre, payable by four annual instalments. The conditions are—(1) Personal residence of the settler, or his family or agent; (2) improvements to be made of the value of 10s. 6d. an acre. Lands from 160 to 640 acres may also be bought at from $1 to $5 an acre, according t ι class, without conditions of residence or improvements.

The Esquimalt and Nanaimo Railway Syndicate have not yet fully arranged the terms upon which they will dispose of their unoccupied lands. They own about 1,500,000 acres, but they are much broken up by rock and mountains.

The land belonging to the Dominion Government lies within the " Railway Belt," a tract 20 miles wide on each side of the line, which begins near the sea-board, runs through the New Westminster district, and up the Fraser Valley to Lytton; thence it runs up the Thompson River valley, past Kamloops and through Eagle Pass, across the northern part of Kootenay district to the eastern frontier of British Columbia. The country is laid out in townships in the same way as in Manitoba and the North-West Territories. The lands may be purchased at a price not less than $5 (£1) per acre—free from settlement conditions, no sale, except in special cases, to exceed 640 acres to any one person. The lands may be "homesteaded" in certain proclaimed districts by settlers who intend to reside on them. A registration fee of $10 (£2) is charged at the time of application. Six months is allowed in which to take possession, and at the end of three years, on proof of continuous residence of not less than six months annually and cultivation, he acquires a patent on payment of $1 per acre for the land. In case of illness, or of necessary absence from the homestead during the three years, additional time will be granted to the settler to conform to the Government regulations. Any person after 12 months' residence on his homestead, and cultivation of 30 acres, may obtain a patent on payment of $2.50 (10s.) per acre. These conditions apply to agricultural lands.

Lands for Sale. In addition to the free-grant lands available in Manitoba and the North-West Territories, several companies have large blocks of land which they offer for disposal at reasonable rates, from $2.50 up to $10 per acre. Among others, the Canadian Pacific Railway Company (Land Commissioner, Mr. L. A. Hamilton, Winnipeg) has about 14 millions of acres; and the Hudson Bay Company (Chief Commissioner, Mr. C. C. Chipman, Winnipeg) has also a considerable area. The same remark applies to the Canada North-West Land Company (Land Commissioner, Mr. W. B. Scarth, Winnipeg) and the Manitoba and North-Western Railway Company; and there are several other companies, including the Land Corporation of Canada. The Alberta Coal and Railway Company also own nearly half a million acres of land in the district of Alberta. The prices of these lands vary according to position, but in most cases the terms of purchase are easy, and arranged in annual instalments, spread over a number of years. Mr. R. Seeman, c/o The Manitoba and North-Western Railway Company, Winnipeg, has purchased about 80,000 acres of land from that railway company. He is prepared to sell the land at a reasonable rate per acre, a small sum being paid down, the remainder in annual instalments on a graduated scale. Mr. Seeman has already sold about 40,000 acres during the last year. As will be seen from some of the delegates' Reports, Lord Brassey, Senator Sanford, and others have land for sale. The Colonisation Board have also land for disposal, under favourable arrangements, particulars of which may be obtained of Mr. G. B. Borradaile, Winnipeg.

Improved Farms. In all the provinces improved farms may be purchased at reasonable prices—that is, farms on which buildings have been erected and a portion of the land cultivated. The following are the average prices in the different provinces, the prices being regulated by the position of the farms, the nature and extent of the buildings, and contiguity to towns and railways:—Prince Edward Island, from £4 to £7 per acre; Nova Scotia, New Brunswick, and Quebec, from £2 to £10; Ontario, from £2 to £20; Manitoba and the North-West Territories, from £1 to £10; and British Columbia, from £2 to £20. These farms become vacant for the reasons which are explained with accuracy in many of the accompanying Reports. They are most suitable for persons possessed of some means, who desire more of the social surroundings than can be obtained in those parts of the various provinces in which Government lands are still available for occupation and settlement.

Agricultural Exports. Canada has already assumed an important position as an agricultural country, and the value of its exports of such products alone now nearly reaches $50,000,000* annually, in addition to the immense quantity required for home consumption. The principal items of farm and dairy produce exported in 1892 — the latest returns available — were : Horned cattle, $7,748,949; horses, $1,354,027; sheep, $1,385,146; butter, $1,056,058; cheese, $11,652,412; eggs, $1,019,798; flour, $1,784,413; green fruit, $1,444,883; barley, $2,613,363; pease, $3,450,534; wheat, $6,949,851; potatoes, $294,421. Besides the articles specially enumerated, a considerable export trade was done in bacon and hams, beef, lard, mutton, pork, poultry, and other meats, as well as in beans, Indian corn, oats, malt, oatmeal, flour-meal, bran, fruits, and tomatoes. The chief importers of Canadian produce at the present time are Great Britain and the United States, but an endeavour is being made, and so far with success, to extend the trade with the mother country, and to open up new markets in other parts of the world. The products of the fisheries, the mines, and the forests are also exported to a large annual value ; and the manufacturing industry is a most important and increasing one, especially in the eastern provinces, and includes almost every article that can be mentioned. It is not necessary to extend this preface or to summarise the

---

* The exports of these products in 1879 were only 33¾ million dollars, and the importance of the present volume of the trade may be realised when it is remembered that prices have declined, roughly, 25 per cent. in the interval.

various Reports; they must be allowed to speak for themselves. They deal with Canada as seen by practical agriculturists, and refer not only to its advantages, but to its disadvantages, for no country is without the latter in some shape or form. It may safely be said, however, that Canada has fewer drawbacks than many other parts of the world; and this is borne out by the favourable opinions that are generally expressed by the Delegation. Those who read the Reports of the farmers who visited Canada in 1879 and 1880 will realise that immense progress has been made since that time—when the vast region west of Winnipeg was only accessible by railway for a short distance, and direct communication with Eastern Canada, through British territory, was not complete. Considerable development has also taken place since 1890—when the previous Delegation visited the country.

The Canadian Government, in inviting the Delegation, wished to place, before the public, information of a reliable and independent character as to the prospects the Dominion offers for the settlement of persons desiring to engage in agricultural pursuits, and it is believed that its efforts will be as much appreciated now as they were on previous occasions. In Great Britain and Ireland the area of available land is limited, and there is a large and ever-increasing population; while at the same time Canada has only a population of about 5,000,000, and hundreds of millions of acres of the most fertile land in the world, simply waiting for people to cultivate it, capable of yielding in abundance all the products of a temperate climate for the good of mankind. It only remains to be said that any persons, of the classes to whom Canada presents so many opportunities, who decide to remove their homes to the Dominion, will receive a warm welcome in any part of the country, and will at once realise that they are not strangers in a strange land, but among fellow British subjects, with the same language, customs, and loyalty to the Sovereign, which obtain in the Old Country.

For general information about Canada, advice to intending Emigrants, and a description of the Canadian Agricultural and Dairy Exhibits at Chicago, see Appendices (pages 27 to 38).

In addition to the Reports of the Delegates referred to above, the Reports of Professor Long, the well-known Agricultural Expert, and of Professor Wallace (Professor of Agriculture and Rural Economy), of Edinburgh University—who visited Canada in 1893—are also available for distribution, and may be procured from any of the Agents of the Government.

THE REPORT OF MR. JEROME J. GUIRY,

Peppardstown, Fethard, Clonmel.

HAVING been honoured by an invitation from the High Commissioner for Canada, Sir Charles Tupper, Bart., to proceed to Canada and make a report on the agricultural resources of that country, I accepted the invitation without hesitation, but fully alive to the great responsibility I was undertaking.

The Voyage.   I left my home in Tipperary on the 17th August, and went on board the "Parisian" at Moville the following day. The weather being all that could be desired, I had a most enjoyable voyage, and on the fifth day we passed Belle Isle and entered the great St. Lawrence River. As some are deterred from emigrating in consequence of the voyage to America, it may not be out of place for me to state that, as far as I could judge, with very few drawbacks, the voyage out is enjoyable, and friendships are formed on

CANADIAN PACIFIC RAILWAY HOTEL, QUEBEC.

board which are both useful and remunerative in the New Country. The food on board is good, and the servants do all in their power to provide for the comfort of the passengers. Sailing up the St. Lawrence, the views are splendid : to the north the Laurentian range rises, and at their base may be seen the pretty villages of the French-Canadian settlers; on the south bank may be seen the long, narrow strips of land, which appear well cultivated, and quite close to the river stand the houses, built of timber and brightly painted. There is no prettier sight in Canada than the sail up the St. Lawrence.

Quebec.   Having arrived at Quebec, we had time to have a look round the oldest town in America, and, having taken a carriage, with two other fellow-passengers we drove to see the Plains of Abraham, where Wolfe fell; then through the old streets of the town, and, joining our ship, left at daybreak. Passing further French-Canadian settlements, we arrived the following day at Montreal about noon.

The French Canadians.   It may be interesting to say a few words about the French-Canadian settlers, whom one sees from the moment the St. Lawrence is entered. They appear happy and contented. They have their churches and schools every six or seven miles apart; and, though I did not give particular attention to this part of my trip, I may be permitted to say that the French settlers in Canada are a credit to it. They are hard-working, thrifty, and well-to-do; the only fault urged against them, and one which I am bound to say they possess, is that they are devoid of ambition, and that they are so contented and happy that they do not mix much with the outer world, and are content with the uneventful farmer's life. Certainly there are no more loyal subjects to the Crown than the French Canadians.

Ottawa.   Having spent Sunday viewing some of the churches at Montreal, our party left on Monday for Ottawa. A word or two may not be out of place about Montreal. It is one of the finest towns in America. Business just now, considering the depression in the States and elsewhere, is very fair. The first thing that strikes one on arrival at Montreal is its resemblance to a thriving French town; it has all the appearance of a Continental town, with the busy air of an English city. The country around is exceedingly pretty, and the view from Mount Royal is magnificent. The town appears at your feet; you can see every street: and at the distance can be seen the Lachine Rapids. Arriving at Ottawa, we were met by Mr. Hall, of the Department of the Interior, the Minister and the Deputy-Minister of that department not being in town. Mr. Hall wished us to see the Central Experimental Farm for the Dominion. We found it situated about two miles from the city; it contains about 500 acres, and is under the control of Professor Saunders. I cannot express my appreciation of all I saw at the farm. On it all kinds of experiments and trials of seeds take place; all records of the trials are kept and published. All seeds are tested for farmers, free of charge and postage. In no other country in the world is such interest taken by the Government in the farmer's interest; and I must say that this interest is appreciated by the people, as shown by the fact that last year the number of questions asked, advice required, and samples tested was enormous. I saw all kinds of grain—wheat, oats, maize, barley, &c.—besides all kinds of roots, vegetables, plants, and trees, grown with and without all kinds of natural and artificial manures.

EXPERIMENTAL FARM, OTTAWA.

Ottawa to Winnipeg. Having been shown over the Government Buildings and the Houses of Parliament, and after learning something of the government of the country—which I will refer to later on — our party left Ottawa and Eastern Canada for Winnipeg and the West. After a two days' journey in the train, I arrived in Winnipeg at eight o'clock on a beautiful September morning. The country from Ottawa to Winnipeg may be described for the greater part of the distance as a dense forest; but in portions of it, such as the district of Algoma, considerable areas of land are suitable for agricultural purposes. The railway skirts the shores of Lake Superior, and this part of the route is beautifully picturesque. As there is very little land open for settlement along this route, except in Algoma and some other districts, I will describe it as the paradise of the sportsman and fisherman; but I must not forget to say that in many parts various minerals are abundant.

Winnipeg. When our party arrived in Winnipeg, we called on Mr. Smith, the Commissioner of Dominion Lands. Mr. Smith wished us to see the house provided in Winnipeg for emigrants. We called, and found a comfortable building, with bedrooms and kitchen, where immigrants can rest, and remain till such time as they can be provided with employment. A book is kept where all employers who want servants enter their requirements, and where all people looking for employment may state their trades and callings. The usual rate of wages for unskilled servants is, for men, \$1.75 to \$2.25 a day, or \$25 a month, with board. At the time we called all the servants, who, the day before, wanted places were provided for at the rates of wages above given. Mr. Smith wished us to see the market gardens of Mr. Salter, an Englishman, who grows vegetables for the Winnipeg market. These gardens were exactly the same as one sees around London. The demand is very good, and Mr. Salter, who came to Winnipeg a poor man, appears to have done well.

WINNIPEG.

Growth of Winnipeg.

I do not like to leave the subject of Winnipeg without saying something of it. Twenty years ago Winnipeg was but a village of 290 inhabitants; now it has over 30,000, and it grows at the rate of from 2,000 to 3,000 a year. It is bound to be one of the largest and most prosperous cities in Canada, as it is the junction of several railways; and hence I should advise any person wishing to settle near a growing city, and to have his property advance in price, to look out for a suitable location in the neighbourhood of Winnipeg. Good land six miles from the city can be bought at from \$5 to \$10 an acre.

Portage-la-Prairie.

In company with Mr. Booth Waddington, an English delegate, I left Winnipeg next morning for Portage-la-Prairie. In this district is some of the best wheat-growing land in Canada. The objection I have to it is that people here go in almost exclusively for wheat-growing. Some years ago, when wheat sold for a dollar a bushel, this kind of farming did remarkably well; but now that prices have declined, and competition is keener than it was, things are not so prosperous in this locality. Near the town of Portage we passed the farm of a Mr. Snitzer. This gentleman commenced not many years ago with little or no capital, and now he has a good house and place, with 320 acres of land, which all appears to be under crop. We saw the oats and wheat in stook, and they appeared splendid.

Sanford Ranch. Having been invited by Senator Sanford to see his ranch, and to look over the country around Westbourne, where the ranch is, we drove from Portage-la-Prairie, accompanied by Mr. Riley, Senator Sanford's manager, to Westbourne. On the way we called at the house of Mr. Duncan McLeod. We found that the threshing had been done on the farm. Mr. McLeod told us that he bought his quarter-section of 160 acres some years ago at $2½ an acre, and that he can now sell it for $8 an acre. We arrived that evening at the ranch home of Senator Sanford. Having had a hasty meal, Mr. Riley drove us over the ranch ; it consists of deep black soil, suitable for either grazing or tillage. The Senator has over 50,000 acres for sale, and is asking $8 an acre for it. The cattle and horses are very good indeed, and, considering its proximity to Winnipeg, I consider any settler might do worse than settle in this locality. The hay lands running north of Westbourne are prolific, and hay can be cut and put up in the rick for $1½ per ton.

Russell. Leaving early next morning, Mr. Riley drove us through a fine tract of country for 10 miles or so, and dropped in at the wayside station of Woodside, on the Manitoba and North-Western Railway. He was most kind in pointing out the country to us.

I was very sorry to have missed seeing Mr. Lynch's herd of Shorthorns. Mr. Lynch is an Irishman who resides near Westbourne, and I am told his herd is famous in this locality. He goes in for stockfarming, with the result that he has done very well indeed. Arriving late on Saturday night at Binscarth, we started for Russell, a town some 10 miles distant. We spent a quiet Sunday there—a rest we much needed after so much travelling. On Monday morning we made an early start to see the country, and first visited the farm of Mr. Cursitor. His history is this : Ten years ago he came to this locality and homesteaded 160 acres; he soon after pre-empted 160 more; he has now under crop 100 acres of wheat, has 6 cows, a number of pigs, and he is fully supplied with all kinds of agricultural implements. I asked him what his capital was at starting, and he said one waggon and a pair of horses. He must now be worth over $2,000. Dr. Barnardo's Home being quite close, we paid it a visit. We were courteously received by the manager, and inspected the fine herd of Shorthorn dairy cows—56 in all. We were informed that the cattle receive nothing more than hay in the winter, and the grass on the prairie in the summer; they were in grand condition, showing what cattle can do on such pasture.

A Successful Farmer. We next visited the farm of Mr. Setter, and his case presented to me more than any others the advantage of settling in this locality. Eleven years ago he came here, $80 in debt, with a wife, child, and waggon and horses. He homesteaded 160 acres, and now has 280 acres of wheat and 140 acres of oats. He has laid aside some money, and has built

himself a comfortable house in the town of Russell, so as to educate
his children.   His property and stock are now estimated to be worth

FARM SCENE, MANITOBA.

over £1,200, or $6,000.   After an early dinner, we left for Asessipi,
a village beautifully situated in a valley on the Shell River.   Here
we met a Mr. Gill, a Leicestershire gentleman, who works a lumber
and corn mill.   Mr. Gill took us to see some cattle, the property of
a Mr. J. Smith, and I never before saw such beasts.   They were three-
year-old bullocks; live weight, 19 cwt.; and on inquiry I found they
had no other feeding but the grass, which appeared to me to be quite
dried up and white.   I could not believe that the beasts were fed on
this till I was informed that it was the celebrated buffalo grass, which
grows in this locality.   Cattle will go any distance for it, and those
fed on it are said to bear any amount of transport and yet hold their
condition.   Mr. Smith had disposed of a lot of his bullocks for £12
each, or 3 cents the pound, live weight.   After visiting Mr. Gill we
returned to Russell, where we arrived at ten o'clock.   I was anxious
to see someone who came to Canada with money, and I found one
here—an English gentleman who came out a few years ago as a
farm pupil, paying a premium to a farmer to teach him farming (a
system which cannot be too highly condemned, and against which the
Canadian authorities in this country have often issued warnings).   He
brought a considerable sum with him, and, after leaving his farming
tutor, went to ranch in a small way.   He has now a splendid herd of
cattle, mostly of the Shorthorn breed, and has done remarkably

well—so well that, although he has since inherited a large property, he has decided to remain in Canada. It is not often that premium-paying farm pupils succeed so well. Next morning we started early from Russell to see a fine herd of cattle belonging to a man who a few years ago was a labourer in Oxfordshire. He came to Russell, worked as a general servant, saved his wages, and is now the possessor of as fine a herd—75 in number—of Polled Angus and Shorthorns as could be desired. He has, besides that, brought out his father and mother. I asked him if he wished to return, and he said not. The Russell district, in my opinion, is a good one; it possesses the advantages of plenty of wood, good grass land, and abundance of water. Good feeding land can be bought in the neighbourhood of Russell for about $4 an acre.

WHEAT STACKS, MANITOBA.

Neepawa. Leaving Russell on the 13th September at three o'clock in the morning, we arrived at Neepawa (the Indian word for *plenty*) about noon; we took "rigs"—conveyances— and left for Carberry. From the large number of wheat stacks I saw in the immediate vicinity of Neepawa, I thought the word very appropriate, for they were very numerous; I tried to count them several times and failed—the whole country appeared studded with wheat stacks. We arrived at Carberry, which is a small town on the main line of the Canadian Pacific Railway. The drive from Neepawa is a long one, and all the land we passed through appeared taken up, with the exception of the school sections; the Government set aside two sections of land in each township to support the schools. Next morning,

accompanied by Mr. Boyd, M.P., and Mr. Lyon, M.P.P., we went
to see Mr. Hope's farm. Mr. Hope is a Scotchman. He homesteaded his farm 12 years ago, and has succeeded so well that he is
now able to retire, having purchased a quantity of house property in
Carberry. He has also provided for his son; and it has all been made
out of his farm. We then visited his neighbour, Mr. Riddle, who
came out some few years ago from county Limerick. Mr. Riddle
has one of the best-kept farms in the district. He says he has done
well. He looks happy and contented. Mrs. Riddle assured me that
she would like to see old Ireland and her friends once again; but
she said she had nothing to complain of. The same day we visited
the farm of a Mr. McKenzie. We were unfortunate enough not to
find him at home, but one of his boys showed us over his place. He
farms over 2,000 acres, and I noticed what would astonish any farmer
in the Old Country—a large heap of manure in the yard which had not
been cleared away for the past six years. This fact shows the richness
of the soil. Otherwise, Mr. McKenzie would be forced to use manure

GRAIN ELEVATOR, BRANDON.

Experimental Farm, Brandon.   The next day we visited the experimental farm at
Brandon. This is the experimental farm for the
province of Manitoba; almost every province has its
experimental farm. We were received and shown over
the place by Mr. Bedford, the manager. It is hard
to say what was most interesting, but I was much struck by the
making of ensilage. On this farm, the stuff to make the ensilage

with is cut up before being placed in the silo, and this operation is performed with a simplicity and cheapness that cannot be excelled. The power is supplied by a windmill, an air-motor machine made in Chicago. The cost of this would greatly depend on the power required, but the fan and driving-rod of the machine can be had in Chicago for $35, or about £7. If any readers require particulars, Mr. Bedford will be happy to supply them. A sight well worth a visit in Brandon is Christie's saw and lumber mills, where 40,000 ft. of timber is cut daily. The logs to supply the mills float down the river a distance of 900 miles. This will give some of my readers an idea of the extent of the country. There are some good shops in Brandon; and Wilson & Smith's furniture warehouse would do credit to London, or any of the cities in the Old Country. The district around Brandon is mostly used for mixed farming. We spent two days driving round the neighbourhood, and from farmers we heard the repeated tale of prosperity. We brought our guns with us, and from our "rigs" we bagged a large number of prairie chicken.

To Indian Head. The next day was Sunday, and, as we were bound to make connection with the West, we took the express for Indian Head. We passed on the way Messrs. Bouverie & Routledge's large farm at Virden. All the corn appeared to have been stacked. We arrived at Indian Head about five o'clock, and next morning visited the experimental farm, which is managed by Mr. McKay, who showed us over it. It is well kept. In one building is placed the grasses grown from seeds which the manager supposes would suit particular districts. He told us that the grass which he found possessed most nourishment, and the best adapted for laying down in his district, was a new Austrian grass called *Bromus inermus*. This he considers would cut fully 3 tons an English acre, and in his opinion ought to be tried in the Old Country. After leaving the experimental farm, our party drove to the farm of a Mr. Harrop, which is close by. He was threshing, we found. He had a splendid sample of wheat, and expected to thresh over 42 bushels to the acre. We then went to Mr. Dixon's farm, and also found him threshing. He said he should be much surprised if he did not realise 40 bushels to the acre. This is some of the best wheat land in Canada; we found the heaviest crops of that grain here. It may be interesting to my readers to know how it is cultivated in this locality. We will suppose that the land is in stubble. The stubble is cut about 18 in. high: this is done with two objects—first, to hold the snow and keep it from drifting; and, secondly, to provide fuel, as after the snow melts away from the stubble the wind dries up the straw. It is then set on fire, and that leaves a splendid bed for the seed, which is then put in broadcast. A cultivator is then put over it, and so the crop is left. Very simple, some will say; and so it is; and this is the reason—that a man is supposed to handle 120 acres of wheat at all times, but during that part of the harvest he requires two hands. Of course my readers will understand that a farmer will not treat all his land in this fashion; a good farmer will have one-third of his land in summer fallow, so as to rest the land. In

my opinion, the country around Indian Head will give the highest rate
of grain this year, and an average of 32 bushels an acre is expected.
Before leaving this locality, we wished to see the celebrated Bell Farm,
which was at one time the largest wheat farm in the world. This farm
was owned and worked by a company. It has now been divided
into several farms, and only a portion is owned by Major Bell.

CAMERON'S FARM, QU'APPELLE.

Qu'Appelle.   We left next day for Qu'Appelle Station, and on the
way we called at Mr. Fraser's farm. The owner told
us he started with little or no capital. He has worked
his way so well that he has over $1,000 worth of machinery, and a
good balance at his banker's. He keeps his books with a surprising
degree of accuracy; and, taking the past five years as an average, his
profits, after all expenses, and feeding himself and family, are $3.80, or
nearly 16s , an acre. Staying that night at Qu'Appelle Station, we started
early next morning for Qu'Appelle Fort. On the way we called on
Mr. Cameron, a Scotch gentleman, who manages a large farm
owned by the Messrs. Sykes. He spoke well of the district. Messrs.
Sykes own 15,000 acres of good wheat land, and we were informed
that they are selling some on the deferred payment system. The country
around appears to be a very good one for wheat-growing. Having
arrived at Qu'Appelle Fort, we visited the schools kept by the Oblate
Fathers for the training and education of young Indians. We were
received and shown over the establishment by Father Hugeneaux.
Nothing could exceed the neatness and cleanliness of the establishment.
It is not long established, but it has done a deal of good. On the
following morning I visited the new roller mills of Messrs. Joyner &
Elkington. Mr. Elkington is an Englishman, who has put his capital
in this mill. He has done well, and speaks highly of this district. I

saw the wheat sold at his mill, and for quality it cannot be excelled. I cannot leave the subject of Qu'Appelle without referring to its beautiful situation. It stands between two lakes, and the hills rise up on each side. I thought it a pity it was not called "Killarney:" to me it appeared one of the prettiest spots in Canada. It has one drawback— one which may soon be remedied—that is, want of railway communication. The country is suitable for cattle-farming.

HYDE FARM, QU'APPELLE.

Regina.   We left Qu'Appelle for Regina, which is not far distant, and arrived in a few hours. Regina is the capital of the North-West Territory. I was glad indeed to meet here a Tipperary man—one of the most brilliant and talented of my countrymen—I mean Mr. N. Flood Davin, M.P. Mr. Davin has done a good deal for the farmers, and they appear to appreciate his services, for there is no more popular member of the Dominion Parliament. There is one striking fact about Regina—that, whatever advantages Nature has denied it, man's energy has made up for the loss, for evidences of push and energy are everywhere apparent.

Prince Albert.   Leaving Regina on the 22nd September, at night, we arrived in Prince Albert the following morning in time for breakfast. Prince Albert is beautifully situated on the river Saskatchewan. It is now a small town, but is rapidly increasing. The country around is picturesque. The north side of the river is very sparsely populated—only a few half-breeds; but on the south side are to be found some creditable farms—notably

Mr. Thomas McKay's, from which there is a grand view. Mr. Thomas O. Davis, a merchant in Prince Albert, has a good property, which he farms; he keeps about 200 head of cattle, some pigs, and about 50 acres of tillage; all this he does with three hands. I do not know of a nicer district in Canada than the country around Prince Albert. Driving out of the town to the south, you pass by a number of small lakes; they literally swarm with wild duck, and the prairie abounds with chicken. Along the lakes are a number of "sloughs," which produce splendid crops of hay. Prince Albert is certainly a favoured district, provided with everything that a farmer may want. At present it is situated in a remote district. It has railway connection with Regina, and other lines may be built shortly; and when the proposed railway is built to Hudson Bay, Prince Albert will be as near England for at least four months of the year as Montreal is now. I was pleased to find a countryman of mine here in the person of Judge Maguire, the Judge of the Superior Court. The Judge is one of my countrymen whom I am proud of—a sterling, good Irishman. Near Prince Albert are situated the large lumber mills of Messrs. Moore & MacDowell. Mr. MacDowell is one of the most prominent men in the West; he has a fine residence, and his partner is an Irishman. Major Moore fought in the Riel Rebellion, and unfortunately lost his leg. The firm employ a large number of hands, and supply the surrounding district with cut timber. The only thing I ever heard urged against the Prince Albert district is, that it is cold in winter; be that as it may, I have no hesitation in saying that Prince Albert is one of the best districts in the North-West. Between Prince Albert and Regina the country is studded with small lakes, good pasture, with plenty of shelter, and splendid hay land. I regret very much that time did not allow me to see the neighbourhood of Duck Lake. I am told it is equal to Prince Albert, but I only saw it from the train. I had the pleasure of a short interview in the train with Mr. Hillyard Mitchell, an English gentleman who goes in extensively for stock-farming. I should strongly advise any young man with, say, £400 capital, to try this part. Mr. H. Mitchell will be happy to supply all further information. Our party left Prince Albert on the night of the 27th September, and arrived in Regina the following morning.

British Columbia. That evening we started for British Columbia on board the West-bound train. About noon the next day we got our first view of the Rockies, and about three o'clock we got to them.

There are a few nice ranches between Calgary and the Rockies, a distance of about 70 or 80 miles. Time did not allow us to visit them. Leaving the open prairie behind, and having attained an altitude of over 3,000 ft., the line enters a gap, and 50 miles further shows an ascent of 1,000 ft. A description of the Rockies would scarcely be fitting to an agricultural report. I will only say that I do not think pen ever wrote aught that would convey anything like the grandeur and awfulness of the reality. Two days after leaving Regina we arrived at Vancouver—a town the population of which was only 1,000 in 1887; in 1888 it was 6,000, and in 1890 15,000; it now exceeds 18,000.

Sunday, the first day we spent in Vancouver, was wet—as, indeed, were most days that we spent in British Columbia—and we were forced to remain indoors all day.

New Westminster. On Monday we proceeded to New Westminster, a town of about 9,000 inhabitants, about twelve miles distant. The town is situated on the Fraser River, and is the centre of the salmon curing and canning industry; how many canneries there are in the locality I could not make out, but I will describe one that I visited. They are large wooden buildings situated on the bank of the Fraser; the proprietors buy the fish from anyone who brings them, and the prices range, according to the quality of the fish, from 6 to 20 cents each. Some idea of the quantity of fish tinned on the Fraser River may be realised when I mention that one cannery I visited turned out 20,000 cases, of 48 tins to the case; the entire money value of the fish taken on the Fraser River this year falls little short of £180,000. The Mayor of New Westminster engaged a steam launch for our party, and, accompanied by this gentleman, the Mayor of Vancouver, and some of the leading citizens, we proceeded up the Fraser River, and arrived at Ladner's Landing about two o'clock. Here we were received by Mr. Ladner, who had "rigs" in readiness for us. He conducted us over a fine part of the country. He pointed out to me some lands in the vicinity of the village which he improved, and has now let at $10 an acre. This appeared to me exactly like some of the land in the county Meath—good fallowing land. Driving to the south, we passed through some heavy low-lying land, fenced in, with very good houses on it. It looked like our best hay land in Ireland, and, on inquiry, I found that it was used for hay-growing purposes. It is in the hands of a land company, and is for sale; $60 an acre is asked for it. To anyone wishing for a climate like the South of Ireland—damp and warm—and who wishes to avoid all trouble of building and fencing, I should recommend this part of British Columbia. I thought a drawback to the district was the bad roads; and I also thought that the lands required to have cuts made so as to remove more of the surface water. Lulu Island is situated in this district, and consists of deep soil, all alluvial, and of the best description. There is no land open to homesteading; it is all in the hands of private owners, who are selling for from $50 to $70 an acre. This district I would consider best for dairy farming; it is too heavy for grain-growing.

Victoria. I regret very much indeed that time did not permit us to see more of what are known as the "delta lands;" but, as we were obliged to push on, we left Vancouver for Victoria, the capital of British Columbia, which is situated on the Island of Vancouver. After a pleasant voyage of 10 hours, we arrived there. Next day, accompanied by the Honourable Mr. Beaven, we started at an early hour to see as much as time would permit of Vancouver Island. We first visited the farm of Mr. Henry King. Mr. King was the only tenant farmer I had so far come across. He holds his land at 12s. an English acre, and he makes his rent by selling vegetables at about double the price he could get for them in Covent Garden. He also keeps a dairy, and disposes of his milk at 30 cents the gallon; at that price it is easy for him to pay rent. We then visited the property of Mr. Jas. Nicholson. He came to Canada some 30 years ago from Kerry. He spoke very highly of the country, and said he had all he desired. I asked him about the prices he obtained for his produce, and he instanced the fact that in some months of the year he was able to get as much as 5s. for a single duck. We drove further on, and saw several plots of land which had been cleared by English families with a view to fruit-growing. The timber in this district is of immense size, and the cost of clearing an acre, we were informed, is about $100 (or £20), but when cleared, the produce from an acre is immense. Returning to Victoria, the Mayor had us shown over the Chinese quarter of the city. It was most interesting. The Chinese live by themselves; they do all the servile work, all the washing, and all the small jobs; they are excellent cooks and indoor servants. Their pay is about $1 a day; it is asserted that without them work could not be done in British Columbia, wages being so high. Mechanics receive very high wages, and two masons told me that they could afford to come back to England some winters and return in April again; they told me that they received as high as $5 a day. The demand for such men is not, however, large, and it is met by the supply on the spot.

Mission City.   Bidding good-bye to our Victoria friends, we left for Vancouver on our return home. Having arrived there, we drove direct to the terminus of the Canadian Pacific line, and caught the East-bound train. We alighted at Mission City (so called from the Roman Catholic Mission to Indians), and were met by all the English settlers in the locality. They divided our party and drove us over a low-lying, wet, but remarkably rich country, full of splendid grass, and well adapted for dairy farming. We visited the pig ranch of Mr. Page, and found 200 pigs being fattened by the simple process of giving them lots of crushed grain and allowing them afterwards to drink water. Mr. Page carries on a large dairy, and is doing well. His work is carried on by Chinamen, to whom he pays $1 a day. The Roman Catholic Mission have a very large property here, which they hold for sale at prices they cannot now obtain. The farmers around Mission City, at the request of Mr. French, had a large collection of fruit and vegetables on view for us at the office of the Board of Trade. This collection was the best I had seen in Canada. I must mention a circumstance in connection with this show. A workman, hearing me talk of trout-fishing, asked me if I would like to see what he could produce. He took his fishing-rod, and, without going further than the bounds of the town, returned in an hour with a basket of trout—splendid fish, one weighing 3½ lbs., and another 2 lbs.

The people of Mission City appear to me to have plenty of enterprise. They have erected a large salmon cannery, and will now give it to anyone to work for three years, free of cost, rent, and taxes. After a pleasant visit we left for Calgary. Allow me to say a few words about British Columbia. In my opinion, it is a grand province. It is as yet in its infancy, and there is no doubt that there is a great future before it. It is bound to be the home of many an Irishman. As to its climate, it is somewhat like Ireland, or the South-West of England. There is little frost, and in the valleys the snow never stays for any time. Everything will grow there, and a farm of 100 acres in this province is worth three times as much as in the East. The prices to be obtained for produce exceed the London prices, and I have seen potatoes sold in the market in September for 1d. a lb. Of course I only write of the province from a week's view, and in that week only a fringe of it was seen. There are hundreds of thousands of acres of fertile land in British Columbia.

CALGARY.

Calgary. Calgary is the headquarters of ranching in Canada; it appears to be situated at the foot of the Rockies, but they are nearly 80 miles distant. It is built on a beautiful, bright river called the Bow, on a lovely plain, and is a substantial stone-built city, with shops quite as good as those in Grafton Street in Dublin; it has all the appearance of prosperity. In Calgary are the land and emigration offices, and the junction of the Red Deer and Edmonton Railway to the north, and of that to Fort McLeod on the south. The day after our arrival in Calgary we were taken by Mr. Rowe, the land agent in Calgary, to see the ranch of Mr. R. G. Robinson. It consists of three sections of land which he owns; he also leases a large quantity from the Government. He puts up a large quantity of hay every year, and says that without doing so ranching on a large scale is very risky. He keeps 300 horses, and 1,300 head of cattle. He breeds his own cattle and horses, and every year raises 300 calves. He goes in entirely for the Hereford breed; every year he sells about 300 beasts, which weigh about 8 cwt. each, and the price he obtains for each is from $40 to $50. I asked him how he did with

his horses, and he said not at all well lately. Some years ago he used
to get for a pair of trained three-year-olds about $350; last year the
price fell to $250, and he does not think it will be as high this year.
Mr. Robinson trains his horses himself, and this year, with the help of
a negro rough-rider, he broke in 40 horses to the saddle and harness,
and in no single instance did he meet a horse that he did not make a
success of.   I am often asked about ranching.   There are, in my
opinion, two kinds of ranching.   One I will call the ranching on a
large scale, generally carried on by partners, or by companies, in this
way : A large quantity of land is rented from Government, on which
a house and extensive buildings are erected; the proprietors keep a
large number of cattle and horses, and, as a rule, do not put up any
quantity of hay—they depend on the soft winds from the Pacific (which
are known as the "chinook" winds) melting the snow; they succeed
for years, but a bad winter may come and they may make heavy losses.
This system of ranching requires a large capital, and is mostly carried
on to the south of Calgary, in the McLeod district.

RANCH SCENE, ALBERTA

Small Ranches.   The other kind of ranching is one I should recommend to any of my friends, and is very like the life
enjoyed by a gentleman farmer in the West of Ireland.
It is this: A homestead is taken ; on this homestead a small quantity
of grain of all kinds is grown. The homestead is so selected that
there is good hay land and plenty of water in the district. A small
house, called a "shack," is built, with a barn, stable, and small out-
offices ; they are built by degrees. The rancher commences with
from 60 to 100 young beasts. Each year in August he puts up
about 3 tons of hay to the beast. This hay grows wild, and has

not to be cultivated, as some of my readers will suppose; it is most excellent stuff. The process of saving consists of cutting it with a machine to-day and placing it in a rick the next day: the climate is so dry that no exposure is necessary. Then, when the winter comes, the rancher gives out his hay to his beasts in the same way as it is given out in Ireland—that is, in a sheltered position. It speaks volumes when I say that after the winter the cattle, though not put under shelter, are, if they are four years old, fit for the butcher. Some few, wishing to have them extra good, give them grain—a system that I cannot too highly advise as the best way of disposing of the produce of the ranch. This latter kind of ranching does not require a large capital: from £500 to £800 is quite sufficient. It is by no means a very hard life; it is free from care, and to the sportsman a most enjoyable means of existence.

To Edmonton. I will now give a description of the journey from Calgary to Edmonton, which I might with truth call the most interesting part of my trip in Canada from an agricultural point of view. From a long experience of judging land in Ireland I ought to know something about it, and, in my opinion, this is one of the best, if not the best district in Canada. Going north from Calgary to Olds, the country is a rich, flat, treeless prairie. It appears to be admirably suited for sheep, and so I am informed it is. There are few settlers on it, and I understand it is leased by ranchers for sheep-farming. At Olds the prospect changes, and the traveller finds himself entering a rich country, studded with fine pieces of bluff ("bluff" is the name given to small natural groves of poplar), natural meadows, and swift clear rivers which abound in fish. This kind of country continues to Abbeyfaile, and on to Red Deer. Our party stayed at Red Deer for three days, and were charmed with the district. A large portion of the Red Deer district—about 60,000 acres—is owned by a land company in Toronto. The company have held the property for many years, and have only now placed it on the market. It is offered on what appear reasonable terms—from $3 to $8 an acre—and the price is payable in yearly instalments over 10 years, with interest at 6 per cent. While we stayed at Red Deer, the third annual fair was held (they call agricultural shows in Canada "fairs"). When one considers the short time this part of Canada has been open for settlement, the exhibits were wonderful. The cattle were fine—they were mostly of the Hereford breed; the wheat excellent; and the oats were the best I ever saw—they were what are known in Ireland as potato oats. We were told that they weighed over 50 lbs. to the bushel. This we did not believe, till Mr. Fraser, the Scotch member of our party, weighed them, and found them to weigh 51 lbs. to the bushel. What pleased us most at the fair was the ladies' department—bread, knitting, &c., &c. The bread was some of the best I ever ate, and the show of butter would do credit to Cork or Limerick. The prize for butter was given to Mr. Trimble, of Twin Pine Creek. During our stay at Red Deer, we were taken to see the country north-west and west of Red Deer. It is well adapted for ranching in the small way I described—full of good hay land, plenty of wood and water. I can recommend this district. The second day we stayed at Red Deer we went to see Mr. Gaetz's farm. It overlooks two lakes. Mr. Gaetz keeps a dairy of 40 cows, and cultivates about 60 acres; his out-offices and stables are built of wood, and are well arranged. After leaving Mr. Gaetz, we were driven to the east, a part of the country rather rough for cultivation, but very well adapted for ranching, with some fine springs of water.

On our way back to Red Deer we called on Mr. Trimble, the same who took the prize at the agricultural show. Mr. Trimble is the son of an Irishman who settled in Ontario. When living in the East he was troubled with chest complaint and hæmorrhage of the lungs. He was advised to try the North-West as a change. He did so, and settled with his family, taking up a section which he calls Twin Pine Creek. He goes in for dairying, and has made it a great success, getting as much as 1s. 4d. per lb. for his butter. Mr. Trimble says that his life is a new one since he came to the North-West.

Edmonton.   The next day we left for Edmonton, but, as our train did not start till late, we had time to see a colony of Icelanders, who are doing remarkably well. They came out two years ago without a dollar. I am informed they make excellent settlers, being frugal and industrious. We arrived that night at Edmonton, and, as it was very late, stayed at the Station Hotel, which is situated on the south side of the river Saskatchewan.

Next morning we drove out to Mr. McKinnon's farm. He has been settled here for the past seven years, and has done remarkably well. He is the local magistrate. This year he has tried an experiment in planting about an acre of apple trees. I hope he will succeed in making the first orchard in the district.

A great deal has been said and written about the Edmonton district. The town is divided by the river Saskatchewan, and as yet is not connected with a bridge. The railway stops at the south end, and this part of the town appears to be fast improving. On the north side of the river are the Hudson Bay fort and offices, and all the public buildings. Go where you will about Edmonton, the land is first-class. One of the largest seams of coal in Canada is to be found in this locality; it runs from near Red Deer River to Edmonton, varying in thickness. There is no doubt a great future before this district; it has all the attributes to make it a great town. Some 40 miles to the north of Edmonton there are large forests. This ensures a plentiful supply of lumber. I visited many other farms in this locality, but, as space will not allow me to enter into details of each farm, I will only give a few particulars of the yields as I heard them from the farmers at the Edmonton Show, which happened to be going on when we arrived. The wheat, I heard, was good; the oats and barley splendid. One farmer who resided about six miles to the east told me that he had 120 bushels of oats to the acre. From inquiries that I made from reliable persons, I was informed that the average yield this year would be, for oats, 75 bushels to the acre, or—to make it plain to my Irish readers— 13 Irish barrels to the English acre. The quality of the oats was excellent. Our party wished to see the land and the stubble on which the oats were grown. We started the following morning, and visited the farm of Mr. David Holmes. We found Mr. Holmes engaged at cooking his mid-day meal (he is a bachelor), which consisted of prairie chicken and bacon. Having invited us to partake of his hospitality, he showed us his homestead. We saw some of the oats in stook, and it appeared as thick as it could possibly be, and he assured us that the produce of one field would not be much short of 120 bushels to the acre. Passing to the south of this homestead, we came to a Mr. Conscadon's farm. He assured us he was not worth a dollar eight years ago, but was now comfortably off. His yard appeared fully stocked with horses and cattle. We then drove to Mr. Walker's farm. Mr. Walker appears to be one of the most prosperous and popular men in the district. Coming from Ireland some years ago, he first settled in Ontario, but, wishing to be able to provide for a large family, he moved to the North-West. He now farms over 1,200 acres, has his farms well stocked, and has every comfort one would wish for. His son Frank told us that when his father left Ireland he was not worth £10 ; he (the son) rode with us the entire afternoon, pointing out the most desirable places for settlement, and showed us some excellent land yet open to homesteading.

To Calgary and the East. Our party left Edmonton, with regret, the next morning, on the return journey to Calgary; and we had an opportunity of seeing that part of the country near Edmonton we had passed through during the night coming north : it looked attractive, and was well timbered and watered. When our train arrived about 40 miles from Calgary it pulled up to load a large flock of sheep which had been fed on the prairie; they were in splendid condition, and were being shipped to the East. We arrived in Calgary that evening, and shortly afterwards left by the East-bound train for Brandon, arriving there at ten o'clock the following night. The land we passed through was bare prairie, not settled up, and did not look desirable, inasmuch as I could see no wood or water. When at breakfast in the car we passed two bands of antelopes. Having rested the night at Brandon, we started at eight o'clock next morning for Killarney, where we arrived at nine o'clock after a 70 miles' drive. When in Killarney I paid a visit to the newly erected cheese factory of Mr. Alex. Davis. He was not at home when I called, but his manager gave me all the information I required. He told me his factory was now taking the milk from 100 cows, which is delivered once a day, the price paid being 75 cents per 100 lbs. of milk ; every 10 lbs. of milk makes 1 lb. of cheese, which is sold for about fivepence the pound. The farmer makes about £6 a year of each cow, which cannot be considered bad when one calculates that the only cost is the milking and sending the milk to the factory, as he grazes his cattle free on the prairie. We left Killarney next morning for Winnipeg, passing through a rich country. We were surprised to see the quantity of hay made up for the winter ; the district from Killarney to Winnipeg is one of the best hay districts in Manitoba.

Having rested in Winnipeg for a few days, and having
Ontario.   bid adieu to some of my delegate friends who had to
return to England before me, I left by the East-bound
train, and arrived in Ottawa two days afterwards. Having decided
not to sail from Canada before the 12th November, and having
12 days on my hands, Mr. Burgess, the Deputy-Minister of
the Interior, suggested that I should see something of Ontario.
Accordingly, accompanied by Mr. Smith, of Yorkshire, one of the
English delegates, I left Ottawa for Toronto on the morning of the
31st of October. I was very much struck with Toronto. It is a beautiful city; the streets are well kept, and, like Paris, mostly planted; and it
has all the modern improvements, such as electric cars, electric light,

AN ONTARIO FARM

and telephone. Having a letter of introduction from an old Tipperary
gentleman—Mr. Millett, of Linnostugh—to Mr. Davis, chairman of the
Dominion Brewery Company, and hearing he had some good cattle and
sheep, we paid him a visit. He received us very kindly, and had us
conducted over his farm at Thorncliffe. I know of no farms in the
Old Country better worth a visit. His sheep were some of the best I
had seen, and his Shorthorn bull, " Morning Light" was a grand beast.
Having spent that night at Toronto, we next morning visited the
Agricultural College at Guelph. We were conducted over the
grounds by the president, Mr. Mills. I cannot speak too highly
of this institution, and I only wish space allowed me to enter
more minutely into all the particulars connected with this college.
Suffice it to say, that all that is possible to do for the good and

welfare of the students is done here (as well as for the farmers of Ontario). They are housed and well fed, taught all matters connected with agricultural science, and are only charged a nominal fee. As the college is a Provincial Government institution, students from Ontario are required to pay a much lower rate than those from the other provinces. The lads work on the farm, and get paid for the work done; so the college is, in a way, self-supporting. From what I have seen of these agricultural colleges and experimental farms, I look on them as doing a vast deal of good, and I am certain the country is acting wisely in fostering them. When will our Government learn a lesson from them?

London, Ontario.
Having left Guelph at noon, we arrived at London (Ont.) that evening, passing through a fair country, suitable for mixed farming. I was told that any farms for sale in this district are held at high prices. At eight o'clock next morning we started for Niagara, passing through a succession of vineyards and orchards. I was told that the best fruit-growing district lay between Lakes Erie and Ontario, this being the region where that industry received the most attention. Time did not allow us to visit many of the great fruit farms. The vines are supported by wire, and I am told that the yield is from 3 to 6 tons of grapes per acre. Having spent a night at Niagara, and having taken several good views of the world-renowned falls, I left Niagara for Hamilton, where I arrived about noon, and immediately left for Toronto. I was particularly struck with the fine country that lies between Hamilton and Toronto; the soil was deep and brown, the houses very fine, and each farm had a large orchard on it. This, I was told, was the best land in Ontario. The general appearance of the country around was that of an English county in the Midlands; but one was particularly struck with the hurdle fences, and the absence of the thorn hedges of England and Ireland. I asked the price of land in this locality, and I was told it sold as high as $100 an acre. I arrived at Toronto on Saturday night, and, having spent Sunday in that beautiful city, I left the following morning for the purpose of seeing the county of York, so called in consequence of the number of Yorkshiremen who have made it their home. I spent the first night at Whitby, and the following morning drove to see the Hon. Mr. Dryden's farm. Mr. Dryden is the Minister of Agriculture in the local Parliament of Ontario. He appears to practise the doctrine he preaches as regards good farming, for I have never seen a better-kept farm, or better cattle or sheep. His Shorthorn bull was quite as good as the one I saw at Thorncliffe, and some of his Shrop. lambs were as fine as ever I saw. The land was all in plough, but I examined it carefully, and I found it very good deep clay soil, brown colour, and all the appearance of being easily worked. On my way back from his place I inquired the price of land in the locality, and was told the price asked was from $50 to $80 an acre. Here again apple-growing was to the front. I was led to believe that the orchard on each farm paid the rent where land was rented. Leaving Whitby next morning, I journeyed by the north side

of Lake Ontario, passing through a number of towns, till Kingston
was reached; here the land did not appear as good as in the Toronto
and York County district. After leaving Kingston the land got
rougher, and as night set in we got nearer to Montreal, which we
reached at eight o'clock. Resting the following two days in that city,
I embarked on board the Allan line "Parisian," and, after a splendid
passage of eight days, arrived safely in Liverpool—having travelled
over 18,000 miles, 1,500 of which I covered in my drives.

Where to Go.   When I returned to Ireland, the first question asked
me by most persons was, "What is the kind of land
in Canada?" My answer is, that in writing this
Report I have tried to describe each district; but people, as a rule, like
to have a question answered in a few words, so I must therefore be as
brief as possible in describing generally the land of a country that
is as large as all Europe. As you enter the St. Lawrence you see
mountains on the north side: these are the Laurentians; between
these mountains and the river there is a good flat country, peopled
by French Canadians. On the south side are hills, wooded, and the
land between those hills and the river is settled by the same people.
The land here is all taken up, divided and re-divided, and there is no
room, in my opinion, for fresh British settlements. Between Quebec
and Montreal the land is not good, but the same general remarks
apply. If I might advise, I would say: Go west of Winnipeg, where
there is some of the best soil in the world; it is a rich, deep black
top soil, mostly on retentive subsoil, almost as fine as soot, and
anything planted in it appears to do well. I have seen the fourteenth
crop of wheat without manure. This in a few words describes the top
soil on much of the lands from Winnipeg to the Rockies—about 1,000
miles; but it, of course, varies in parts. The Rocky Mountains are
barren, and are only rich in minerals and timber. Then comes British
Columbia. The general idea is that there is little or no good land in
British Columbia. This is a mistake, as there is a large quantity
of splendid land in the Okanagon Valley; there is also the delta land
near Vancouver, and a good deal elsewhere: this land is a rich brown
land, that will give crops and meadows for many years. This land
reminds me as much as possible of the land banked in from the river
Shannon, called "Corkes land," and which produces the largest hay
crops in Ireland. The land in Ontario is mostly brown heavy top
soil; many parts appear to be over-cropped. I should only recommend
Ontario to those who wish to farm in the same way as in Ireland,
and who wish to enjoy all the comforts of home. Land is, of course,
dear in Ontario, compared with the newer settlements. Speaking
generally, the land in Canada is very good. If it got the same treat-
ment as land gets in Ireland, it would produce double the crop. As I
had heard so much about the cold, I had to find out the exact truth
about it, and I was told almost the same thing by everyone: during
the winter the cold is great, but it is a dry cold; there is little
or no wind, and, of course, no damp. The days are bright; and
everyone is prepared for the winter. Fuel is cheap and plentiful, and
in the town of Winnipeg coal is sold at $5 a ton. Everyone I spoke

to told me they enjoyed the winters. Certainly the cold of winter appears to have very little evil effect on life, as I saw more old people than in Europe, and I believe throat and chest affections are almost unknown.

Laws. The Canadian laws are founded on the English, except in the province of Quebec; there the people, as I have said, are French Canadians, and the old French civil law is in force. This is but right, as the people govern themselves; the laws are their own choosing, as each province makes its own laws, within the terms of the British North America Act; and in each province magistrates, County Court Judges, and Judges of the Higher Courts are appointed, and from their decision there is the right of appeal to the Supreme Court of the Dominion at Ottawa. Paid magistrates, who are lawyers, are appointed in the large towns; but in the country the work is done by unpaid magistrates; and Justices of the Peace have a more extended jurisdiction than at home. Hence it is that very few cases find their way into the High Courts, and a vast amount of money is saved. I saw a case of robbery tried, which did not cost the Government $1, that at home would have cost £50. As to the land laws, they are simplicity itself. As each man is generally the owner of his farm, and as in the West, tenants and landlords are unknown, there is no friction, as in the Old Country. In the Eastern provinces land is let to a tenant, as at home. There is no complication about title to land. It may be asked: How is the government of the country carried on if there be little or no taxation? In this way: The Canadian Government, with a view to fostering home industries, tax imports to a moderate extent; from this source a sufficient income is derived. Canada has no Pension or Civil List to keep up, and, as there are no paupers or beggars, no poor law system is required. As regards the police, there are two branches—one in towns, selected and paid for by the Town Councils; and another in the West, who are mounted, and are a semi-military force: they are appointed, officered, and paid for by Government, and are a splendid force. Crime is almost unknown, and there is an absolute safety of life and property. A question I have often been asked by my farming friends at home is, if there be any danger to be apprehended from the Indians or wild beasts. As regards the Indians, they appear to me to be a quite harmless race. The Canadian Government, to their credit, do all in their power to help them. As to wild animals, they are only to be found in the mountains, far away from settlements. The coyote, a beast larger than our fox, is found in the prairies, and is sometimes dangerous to sheep, but a great coward. As regards game, it is an important subject, inasmuch as it enters into the question of food supply. In each district a person is appointed to look after the game laws. Close seasons are fixed, and the official is responsible for seeing that the regulations are carried out. When the supply of any animal, such as deer, &c., or any kind of birds, is found to be diminishing, shooting is apt to be forbidden for years in some districts. The exportation of game from Manitoba and the North-West is not allowed. Pheasants in British Columbia, prairie chicken and ducks on the prairies, are plentiful. Jack-rabbits, a kind of hare, are abundant. Venison sells in most towns for about 2d. a lb.

Religion. I now come to deal with a very important and delicate subject—religion. In no part of the world is religion in a more satisfactory condition than in Canada. Everyone can go his own way, and few care what religion anyone else professes. There is no State Church; but in Quebec, as I have before stated, the people are Roman Catholic, and they make their own laws, and the payment of tithe is still part of the law in the province, in respect of Roman Catholics. There are also many Roman Catholics in the other provinces, but there they enjoy no special advantages—except separate schools in some places. Everyone can pay or not, just as he likes, to the support of his faith. Somehow or another it has got into the minds of the Irish Catholics who do not know Canada that it is an Orange, Protestant country; how this has got abroad I do not know, but it is an absurd idea, and one that is used largely by Canada's enemies. One noted example of the toleration of the Canadian people will be found in the fact that, out of 14 members of the Canadian Cabinet, the Premier and three more are Catholic. Compare this with the United States, where there has never been a Catholic President, or, I understand, even a member of the Cabinet, professing that faith. I do not like to leave this subject without bearing testimony to the good work done in the North-West by the Oblate Order of Priests and Nuns. They do splendid work in schools, hospitals, &c., and the Government do all in their power to help them.

Railways. The railway system in Canada is as yet in its infancy; but there are about 15,000 miles in operation. They are mostly worked by private companies, but the Government give substantial help in land and money. The postal system is good and regular, and there is a post office in every town and village; there are also book and parcels posts at a cheap rate.

I have been asked as to the social condition of Canada as compared with the Old Country. As far as I could see, there appears to be a great equality of classes; certainly the distinction of caste that prevails so much here has no existence in Canada. This state of things is largely brought about by the fact that, as each man owns his own farm, he does and thinks as he likes. Then in Canada there appear to be very few poor or very few rich families—wealth is more evenly divided; and then everyone works in Canada, and labour has a dignity and respect that it has not in Europe. People in Canada live better than they do in England; in the old parts of the Dominion, and in all the large towns, the hotels are kept pretty much the same as in London; but in the West meals are given at stated times, and cost from 25 to 50 cents each. Meat is used at each meal, hot for breakfast and dinner, and generally cold at supper. The beef is of the very best kind, being fed on grass; mutton is not as good as ours; veal and poultry good and plentiful. Canadians are great consumers of fruit; and tea is served at each meal.

Try the West. As I think I have exhausted all I have to say, and have perhaps tried the patience of my readers, I will conclude by giving my opinion as to who ought and ought not to emigrate to Canada. To those who are well off, and who have every comfort in the Old Country, and who do not wish to commence life over again, I say: Stay where you are; but to the young or middle-aged man who cannot look forward to the future with any degree of confidence, and to those who are weighed down with care, I say: Try the West. There is an opening for any well-conducted man who wishes to work and keep steady. The work need not be very hard; it is steadiness that is required more than hard work. Canada is a good country for a man with energy, who wishes to provide for his family; it is a good country for a man with small or large capital. The want of capital, in my opinion, is keeping back Canada more than any other cause. It is a good country for a man who does not wish to face the distance, the loneliness, and the hardships of the bush in Australia; it is the country for a young man with, say, £400 to £600 capital to run a band of cattle, and to whom an occasional trip to the Old Country will be a pleasure, with the certainty of a splendid return of income and a sure investment of capital. I can also recommend it to those who wish to spend a pleasant autumn trip. Ten times the number of birds can be shot with one-third the cost of a Scotch shooting. I recommend those who have made up their minds to try Canada to be in no hurry in selecting a location when they get there. I met many who are now settled in the East who regret they did not go West. I saw places that I would select at first, but saw better afterwards. A man very often commences an undertaking without a sufficient sum to carry it out; then he gets into debt, and the high rate of interest puts his debt beyond his control. Private money-lenders, and those who sell agricultural machines on credit, should be avoided. I recommend mixed farming in preference to continual and exclusive wheat-growing. The place can be best selected by the settler. Every facility will be given to inspect districts; the Government have in all towns and villages land agents, whose business it is to look after the interests of the immigrant. Besides, farmers are always delighted to see a new man, and are ever ready to give him a hand and sound advice. The reason is, there is room for all. I cannot recommend clerks or shop assistants to go; the same may be said of the professions. Lawyers, in particular, have no opening in Canada; the Canadian Government think that the laws were made for the people, and not the people for the lawyers. The idler has no chance of a living in Canada, and it is no place for the gentleman who is above his business. I have seen men fail in Canada, but they were men who would fail at home. There is a kind of man pointed out as a failure; he is the man who has been sent away, and is supplied with a sum of money as long as he remains. He is known as the "remittance man." When he returns home he is pointed out as a Canadian failure. Such a man is hard on Canada. A steady, industrious man, can make a happy, comfortable home there.

PART VI.   3

I cannot conclude this Report without saying that, in making it, I have tried to avoid any exaggerations; I only give what I saw and heard. The truth can be told of Canada, for it will bear the truth. I wish to thank all that I have come in contact with during my travels for the kindness and hospitality I have received. I wish in particular to mention the High Commissioner, Sir C. Tupper, Bart., who was my fellow-passenger going and returning; also his secretary, Mr. Colmer, for his kindness. When in Canada I received from the Minister of the Interior, the Hon. Mr. Daly, and Mr. Burgess, the Deputy-Minister of that Department, kindness which I shall not soon forget.

A FARM-HOUSE, SOUTHERN MANITOBA.

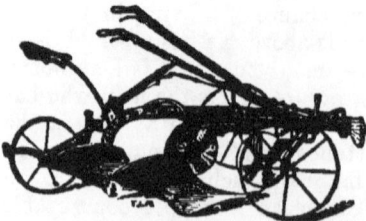

A DOUBLE-FURROW AND A SINGLE-FURROW WHEEL PLOUGH.

APPENDIX A.

GENERAL INFORMATION ABOUT CANADA.

General Information. The Dominion of Canada includes the whole of British North America to the north of the United States, and has an area of nearly 3,500,000 square miles. It is divided into eight separate provinces, and the population at the last census was 4,829,411—viz.: Prince Edward Island, 109,088; Nova Scotia, 450,523; New Brunswick, 321,294; Quebec, 1,488,586; Ontario, 2,112,989; Manitoba, 154,472; the North-West Territories, 67,554; British Columbia, 92,767; and unorganised Territories, 32,168. The extent of the country will be better understood by stating that it is larger than the United States without Alaska, and nearly as large as the whole of Europe.

Constitution and Government. The government of the country has at its head the Governor-General, the representative of Her Majesty. The Dominion Parliament consists of the Senate and of the House of Commons, and the government of the day is in the hands of the majority, from whom the Privy Council, or the Cabinet, who act as the advisers of the Governor-General, are taken. The members of the Senate are nominated for life by the Governor-General, and the duration of the House of Commons is fixed by the Act as five years. Each province has also its local Parliament, in some cases of two Houses, as in Prince Edward Island, Nova Scotia, New Brunswick, and Quebec, and in others of only one, as in Ontario, Manitoba, and British Columbia. The head of the Provincial Government is known as the Lieutenant-Governor, and is appointed by the Governor-General. The constitution of Canada is contained in the British North America Act, 1867, which defines the powers both of the federal and of the local Legislatures. It may be said, generally, that the former deals with matters concerning the community as a whole, and the latter with subjects of local interest. Twenty-seven years' experience has demonstrated that the country has made great progress under the existing institutions, and the principle of union is recognised by all political parties as the sure foundation on which the future of the Dominion depends. There is a free and liberal franchise in operation, both for the Provincial and Dominion Parliaments, which gives most men the benefit of a vote. In the provinces there are county and township councils for regulating local affairs, such as roads, schools, and other municipal purposes, so that the government of the Dominion is decentralised as far as practicable, in the spirit of the Imperial legislation before mentioned.

Climate. Nothing connected with Canada is so much misrepresented and misunderstood as its climate, but it has only to be experienced to be thoroughly appreciated. It is warmer in summer and much colder in winter than in Great Britain; but the heat is favourable to the growth of fruit and the crops, and in every way pleasant and beneficial, and the cold is not prejudicial to health or life. In fact, Canada is one of the healthiest countries in the world. The winter lasts from the end of November or the beginning of December to the end of March or middle of April; spring from April to May; summer from June to September; and autumn from October to the end of November. The seasons vary sometimes, but the above is the average duration of the different periods. The nature of the climate of a country may be measured by its products. In winter most of the trades and manufactures are carried on as usual, and, as regards farming, much the same work is done on a Canadian farm in autumn and in winter as on English, Scotch, or Irish homesteads. Ploughing is not possible, of course, in the depth of winter, but it is done in the autumn and early spring, and in the winter months cattle have to be fed, the dairy attended to, cereals threshed, machinery put in order, buildings repaired, carting done, and wood-cutting, and preparations made for the spring work, so that it is by no means an idle season. One thing is perfectly certain—that the country would not have developed so rapidly as it has done, and the population would not have grown to its present proportions, had the climate been unfavourable to the health, prosperity, and progress of the community. Of course there are good and bad seasons in Canada, as everywhere else, but, taken altogether, the climate will compare very favourably with other countries in similar latitudes.

Temperature.   As the temperature in Manitoba and the North-West Territories is frequently referred to, it is desirable to quote official statistics bearing on the question. The mean temperature at Winnipeg in the summer is 60·3°, and during the winter 1°; Brandon, 58·1° and −1·8°; Rapid City, 62·2° and 2·7°; Portage-la-Prairie, 61·8° and 12·6°. In the North-West Territories, the summer and winter mean temperatures at the specified places are as follows:—Regina, 59·2° and − 2·4°; Calgary, 55·6° and 12·2°; Edmonton, 55·2° and 11·3°. It is very evident the temperature only very occasionally reaches the various extreme limits that are sometimes mentioned, or the mean winter temperatures could not be anything like the figures above quoted.

Products of Canada.   Reference has been made elsewhere to the agricultural products of Canada. The country also possesses great wealth in the timber contained in the immense forests, and in the minerals of all kinds, including coal, gold, silver, iron, copper, &c. Then, again, the fisheries along the extensive coasts, both on the Atlantic side and on the Pacific, and in the inland waters, are most valuable and varied, and are valued annually at several millions sterling. The principal fishes are salmon, trout, cod, herring, mackerel, halibut, and haddock. Oysters and lobsters are also most numerous. The manufacturing industry already occupies a most important position, and is growing more extensive every year. Almost every kind of manufacture is carried on. This activity is not confined to any one part of Canada, but is apparent in all the older provinces, and will no doubt in time extend to the western parts of the Dominion also.

Mortgages. Reference is sometimes made to some Canadian farms being mortgaged. It should be borne in mind, however, that a proportion of the Canadian farmers start with little or no capital. In order to provide capital in such cases, the farm is mortgaged, but the loan companies, as a rule, do not advance more than half the value of the properties. The interest paid bears no comparison to the rent of similar-sized farms in the United Kingdom, and the fact of the existence of a mortgage, in these circumstances, is not detrimental to the position of the farmer. Not only is the interest invariably paid, but the experience is that the loans are paid off as they mature. The losses of the Canadian companies are comparatively small, and the investment, therefore, is a good one to the lender, and an advantage to the farmer.

Trade Imports and Exports. Canada's trade—import and export—amounts to nearly £50,000,000 per annum, and is largely with Great Britain and the United States, the balance being exchanged with European countries, the West Indies, South America, Australasia, China, and Japan. The figures given above only include the outside trade, and do not embrace that done between the various provinces. Free trade, in its entirety, exists within the boundaries of the Dominion, and the local business is very large, as the tonnage carried on the railways and canals and on the coasting vessels will show. It may be stated that the revenue now amounts to about $36,000,000 per annum, of which about $20,000,000—equal to 17s. per head of the population—is obtained from customs duties on goods imported into Canada.

Markets. Markets, either within or without the Dominion, exist for all the products of the country, and no difficulty is found in this respect. New markets have been provided by the establishment of lines of steamers to the West Indies, Australasia, China, and Japan, which are now in operation. Canada is well served with railway and water communication, and the shipping owned in Canada is so large that it occupies a high place in the list of ship-owning countries of the world. A railway extends from the Atlantic to the Pacific Ocean, and each province possesses excellent railway accommodation; in fact, there are over 15,000 miles of line in operation at the present time. The rivers and canals have been so much improved of late years, that the largest ocean-going steamers can moor alongside the wharves at Quebec and Montreal, and it is possible for a vessel of 500 tons burden to pass from the Atlantic into the great lakes. The enlargement of the canals now in progress, which is to be completed in 1895, will permit ocean vessels of 2,000 tons gross burden to pass to the head of Lake Navigation without breaking bulk.

Social Distinctions. The distinctions of class do not exist in Canada to the same extent as in the mother country. There is no law of primogeniture, and there are no paupers; a feeling of healthy independence pervades all classes, which no doubt arises from the fact that every farmer is the owner of his acres, is his own master, and is free to do as he wills—a state of things conducive to a condition of freedom unknown in older countries. Then, again, taxation is comparatively light, and many social difficulties, still under discussion in Great Britain, were grappled with in Canada years ago. Religious liberty prevails; there is practically free and unsectarian education; a free and liberal franchise exists; local option in regard to the liquor traffic is in operation; the duration of the Parliament does not exceed five years, and the members are paid for their services; marriage with a deceased wife's sister has been legalised; and there is no poor law system, although orphans and the helpless and aged. of both sexes are not neglected, being cared for under the municipal system. And, again, a good system of local government is at work in every province. The system of education in force—under the supervision and guidance of the Provincial Governments—enables the best education to be obtained at a trifling cost, and therefore the poor, as well as the rich, can make themselves eligible for the highest positions in the country. In principle the system in operation is the same in the different provinces, although the details may differ somewhat. In each school district trustees are elected to manage the schools for the inhabitants, who pay a small rate towards their support, the balance being met by considerable grants from the local governments.

British subjects settling in Canada do not require to be naturalised. They are entitled to all the same rights and privileges as their fellow British subjects who may have been born there; indeed, the removal of a family to Canada makes no more difference in their position, as British subjects, than if they had gone instead to any city, town. or village in the United Kingdom. Of course it is a different thing if they go to the United States or any other foreign country. In that case they must renounce their birthright, and their allegiance to their sovereign and their flag, before they can enjoy any of the political advantages of citizenship; and in many parts of the United States land cannot be bought, or sold, or transferred, excepting by naturalised persons.

Government Agents in the United Kingdom. Intending settlers in Canada are strongly advised to communicate, either personally or by letter, with the nearest agent of the Canadian Government in Great Britain (see Preface) before they leave, so as to obtain the fullest and latest advice applicable to their cases. Cards of introduction to the Government Agents in Canada are also supplied to desirable persons. Any information supplied by these gentlemen may be thoroughly relied upon.

Then, again, on reaching Canada, or at any time afterwards, the nearest Government Agent should be consulted, as they are in a position to furnish accurate particulars on all matters of interest to the new arrival.

Government Agents in Canada. The Dominion Government has agents at Quebec, Montreal, Halifax, and St. John, the principal ports of landing in Canada; and the various Provincial Governments also supervise immigration as far as possible. At Toronto, Ontario, the Superintendent of Immigration is Mr. D.

Spence, 65, Simcoe Street; and in Victoria, British Columbia, Mr. Jessop, the Provincial Government Agent, should be consulted. The following is a list of the various Immigration Agencies under the supervision of the Department of the Interior:—

Winnipeg, Man. { Commissioner of Dominion Lands, in charge of Outside Service in Manitoba and the North-West Territories } Mr. H. H. SMITH.

Agents at Ports of Call for Steamships in Canada:—

| Mr. E. M. CLAY | ... | Halifax, N.S. | Mr. P. DOYLE ... | ... | Quebec, Q. |
| ,, S. GARDNER | ... | St. John, N.B. | ,, J. HOOLAHAN | ... | Montreal, Q. |

Dominion Lands Agents in Canada who act as Immigration Agents:—

W. H. HIAM ...	...	Brandon, Man.	THOS. ANDERSON	Edmonton, N.W.T.
W. G. PENTLAND		Birtle, ,,	C. E. PHIPPS ...	Oxbow, ,,
JOHN FLESHER	...	Deloraine, ,,	E. BROKOVSKI ...	Battleford, ,,
W. M. HILLIARD...		Minnedosa, ,,	GEO. YOUNG ...	Lethbridge, ,,
W. H. STEVENSON		Regina, N.W.T.	T. B. FERGUSON	Saltcoats, ,,
AMOS ROWE ...	...	Calgary, ,,	JOHN MCKENZIE	New Westminster, B.C.
J. G. JESSUP	...	Red Deer, ,,		
JOHN MCTAGGART		PrinceAlbert,,	E. A. NASH	... Kamloops, B.C.

The best time for persons with little or no capital to go out is from April to July—the earlier the better. Domestic servants may start at any time of the year.

No Assisted Passages. There are no free or assisted passages to Canada. The full ordinary steamship fares must be paid by all immigrants, and they must also have enough money in addition to pay for their railway fares from the port of landing to their destination, and to provide board and lodging until work is secured. The Government does not render any assistance in these matters, and all new-comers must be self-supporting. The Government Agents do not book passengers, and intending emigrants are advised to consult the local steamship agents on that subject. Neither do they recommend any one line more than another. They are quite impartial in both respects.

Booking Passages. It is not necessary to say anything in detail about the various steamers going to Canada, or about the fares. All such information can be obtained from the advertising columns of the newspapers, or from the steamship agents, who are to be found in every town or village. Passengers are recommended to take through tickets (including ocean and rail tickets) to their destinations in Canada. They are issued by the steamship companies, and in this way it is often possible to save money—as through tickets often cost less than the ocean ticket and the Canadian rail ticket if taken separately. Many of the railway companies in Great Britain issue cheap railway tickets from various places to the ports of embarkation, such as Liverpool, London, and Glasgow, and in these cases information may be obtained from the railway booking offices.

Luggage. Passengers should pay particular attention to the labelling of their luggage, and labels may be obtained from the steamship companies. They should also bear in mind that the steamship companies only carry free a limited quantity of baggage, according to the class of ticket taken, and that the railway companies may charge for anything over 150 lbs. weight. The Canadian Pacific Railway carry 300 lbs. free for emigrants proceeding to Manitoba and the North-West Territories. Care should be exercised in deciding what had better be taken to Canada. Furniture, and heavy and bulky goods of that description, had better be left behind, as the freight charged for extra baggage makes it an expensive luxury; all household requirements can be purchased in the country. Agricultural implements and tools should not be taken out, as the most improved articles of this description adapted to the country can be purchased in any village in Canada. Skilled mechanics and artisans, when recommended to go out, may take their tools, but they must remember what is stated above, and also that in the Dominion all these things can be bought at reasonable prices. Emigrants may be safely advised to take a good supply of underclothing, heavy and light, for winter and summer wear, house and table linen, blankets, bed-ticks, and any other articles of special value which do not take up much room.

Settlers' Effects free of Customs Duty. Settlers' effects are admitted free of customs duty if they come within the terms of the following clause of the customs tariff :—

> *Settlers' Effects, viz. :*—Wearing apparel, household furniture, professional books, implements and tools of trade, occupation, or employment, which the settler has had in actual use for at least six months before removal to Canada, musical instruments, domestic sewing machines, live stock, carts, and other vehicles and agricultural implements in use by the settler for at least one year before his removal to Canada, not to include machinery or articles imported for use in any manufacturing establishment or for sale : provided that any dutiable article entered as settlers' effects may not be so entered unless brought with the settler on his first arrival, and shall not be sold or otherwise disposed of without payment of duty until after two years' actual use in Canada ; provided also that, under regulations made by the Minister of Customs, live stock, when imported into Manitoba or the North-West Territory by intending settlers, shall be free, until otherwise ordered by the Governor in Council.

Wages. Wages—which, of course, vary from time to time—are, as a general rule, from a quarter to one-half higher than in Great Britain, but in some trades they are even more. The cost of living is lower, upon the whole, and an average family will, with proper care, be much better off in Canada than at home. There are very good openings in Canada for the classes of persons recommended to go out, but it must be borne in mind that hard work, energy, enterprise, and steadiness of character are as essential there as in any other country. Indeed, perhaps they are more necessary ; but, on the other hand, there is a much better chance of success for any persons possessing these qualities, and who are able and willing to adapt themselves to the conditions of life obtaining in Canada.

Capitalists. Persons with capital to invest will find many openings in Canada. They can engage in agricultural pursuits, taking up the free-grant lands or purchasing the improved farms to be found in every province, or in mining, or in the manufacturing industries. Again, a settled income will be found to go much farther in Canada, and while the climate is healthy and the scenery magnificent, there are abundant opportunities for sport; and the facilities for education are not to be excelled anywhere.

Where to go. Young men should go to Manitoba, the North-West, or British Columbia. Older men, with capital and young families, should go to one of the older provinces, and either buy or rent an improved farm. This, however, is only a general statement, and individual cases must be decided by the special circumstances of each. In Manitoba and the North-West, and in some parts of British Columbia, pioneer life on free grants,or away from railways, is attended with a certain amount of inconvenience, and an absence of those social surroundings which may be obtained in the older settled parts of these and other provinces, and this fact should be borne in mind by those who are considering the subject. But even in these districts improved farms may be purchased at reasonable rates.

Capital necessary. It is difficult to lay down a hard-and-fast rule as to the amount of capital necessary for farm work. The answer depends on the energy, experience, judgment, and enterprise of the person who is to spend the money, the province selected, whether free-grant land is to be taken up or an improved farm rented or purchased, and many other details. It may safely be said, however, that if a man has from £100 to £200 clear on landing, and some knowledge of farming, he is in a position to make a fair start on the free-grant land in Manitoba and the North-West; but it is generally advisable to obtain some experience of the country before commencing on one's own account.

Farm Servants. There is a large and growing demand for male and female farm servants in every part of the Dominion. Machinery of various kinds is in daily use, but labour is scarce notwithstanding, and good hands can in the proper seasons find constant employment. Many persons of this class who started as labourers now have farms of their own in some of the finest parts of the Dominion. Market gardeners, gardeners, and persons understanding the care of horses, cattle, and sheep, may also be advised to go out.

Domestic Servants. So far as numbers are concerned, perhaps the largest demand of all is for female domestic servants. The wages are good, the conditions of service are not irksome, and comfortable homes are assured. Domestic servants should go at once on their arrival to the nearest Government Agent. These gentlemen often have in their offices a list of vacant situations, and will refer applicants to the local ladies' committees, so that they may have the benefit of supervision and guidance, until they are satisfactorily placed. Servants should, however, take their characters with them, and must bear in mind that good records are just as indispensable in Canada as elsewhere. They may safely go out at any time of the year.

Other Classes of Labour. There is little or no demand for females other than domestic servants. Governesses, shop assistants, nurses, &c., should not go out unless proceeding to join friends able and willing to aid them in getting employment. Mechanics, general labourers, and navvies are advised to obtain special information as to their respective trades before going out. The demand is not now so great as it was a few years ago, and such men, especially with large families, are not encouraged to set out *on the chance* of finding employment. They may be safely advised to start when going to join friends who advise them to do so, or if they have the inclination and the knowledge to enable them to change their callings and become agriculturists.

Clerks, draftsmen, shop assistants, and railway *employés* are not advised to emigrate unless proceeding to appointments already assured. Any demand for labour of these kinds is fully met on the spot.

No encouragement is held out to members of the legal and medical and other professions, schoolmasters, and persons desiring to enter the military and civil services, to go to the Dominion, especially in cases where immediate employment is necessary. There are always openings and opportunities for men of exceptional abilities with a little capital; but, generally speaking, the professional and so-called lighter callings in Canada are in very much the same position as they are in the United Kingdom, the local supply being equal to, if not greater than, the demand.

CANOEING.

APPENDIX B.

THE CANADIAN EXHIBITS AT CHICAGO.

The Canadian exhibits at the Chicago Exhibition having been referred to in several of the delegate's Reports, it has been considered desirable to publish such facts as are available as to the success which the Dominion obtained on that occasion in competition with the world. The American Press are unanimous in conceding that Canada will reap a greater benefit from the World's Fair than any other country. The variety of the vegetable products of Canada served to illustrate in a manner, hardly to be shown in any other way, the climate and the fertility of its soil; while the exhibits of mineral wealth, of its fisheries, and of its manufacturing industries demonstrated the possibilities of future development.

It may be said that Canada was unrepresented on many of the juries connected with several of the groups of exhibits, and on others the Canadian members were of course in a minority. It is eminently satisfactory to find, therefore, that the awards in all classes of exhibits have been so numerous, and frequently coupled with remarks of a flattering nature.

The following is an extract from the report of the British Consul at Chicago to the Earl of Rosebery, Secretary of State for Foreign Affairs, on the Chicago Exhibition:—

Canada has been brought prominently forward in a manner which can scarcely fail to assure permanent benefit. Its chief exhibits were natural products, though the colony was represented in every department except electricity. Its cheese and butter exhibits were remarkable, and gained a disproportionately large number of awards, beating all competitors; Japan is understood to have sent a special commission to examine and report on the methods adopted by the colony in these matters. The show of animals, especially sheep, met with great approval. The quality of Canadian fruit was generally recognised. The exhibit of grain and other products of the north-western provinces has shown what can be grown, and as a result many inquiries have been made with a view to settlement in those parts. The same applies to British Columbia, regarding which province overtures have been made by quite a colony of Austrian subjects for settlement, with a view to fruit-growing and general farming.

Agriculture. The Canadian exhibits in this important department were excelled by none in quality and appearance. The excellence of the wheat was the subject of general comment, and a considerable demand has already arisen on the part of United States farmers for seed grain from Manitoba and the North-West Territories. Canada obtained 1,016 awards in this group, including 776 awards for cheese and butter. This does not comprise the awards obtained by Manitoba, which have not yet been received. It is understood that in the tests for quality, made under chemical analysis on behalf of the Exhibition authorities, Manitoba No. 1 Hard Red Fyfe wheat gave the very best results.

The exhibitions of cheese and butter were the largest of
Cheese and   their kind ever made on the North American continent.
Butter.   Two competitions were arranged for Cheddar or factory
cheese, in the months of June and October.   In the
first named, the United States entries numbered 505, and the Canadian
162.   There were 138 prizes awarded, of which Canada took 129, and
the United States 9.   Thirty-one exhibits of Canadian cheese also
scored higher points than the best United States cheese.   In the
October competition for the same class of cheese, made previous to
1893, there were 82 entries from the United States, and 524 from
Canada.   There were 110 prizes offered, and Canada secured them
all.   There were also 414 awards for cheese made in 1893.   Of these,
Canada obtained 369, and the United States 45.   In these two competitions, therefore, the United States entered 587 exhibits and took 54
prizes, as against Canada's 686 entries and 608 prizes.   There were three
judges for cheese, two for the United States, and one for Canada.   The
significance of this result is enhanced when considered in conjunction
with the difference in the population of the two countries—65 millions
against 5 millions.   Canada now exports several millions of pounds of
cheese per annum more than the United States to the English market,
her exports to the mother country having risen from 30,889,353 lbs.
in 1875 to 127,843,632 lbs. in 1892.   In the butter competition,
Canada took 27 awards.   The great development of the cheese industry
in recent years has interfered, no doubt, with the expansion of the
butter trade.   The steps, however, that have been taken of late years
to encourage this industry are having effect; and the Dairy Commissioner of the Dominion has expressed an opinion that within five years'
time the manufacture of butter in Canada will be equal to that of
cheese, both in quality and quantity.   In 1893 Canada exported
43,193 cwts. of butter to Great Britain.

The absence of awards for Canadian agricultural
Agricultural   machinery is explained by the withdrawal of the
Machinery.   exhibits from competition, it having been decided
that machines adapted to field work should be judged
by field tests.   As this would have necessitated bringing duplicate
machines to Chicago at great expense, or the spoiling of the actual
exhibits for show purposes during the remainder of the Fair, the
greater number of Canadian and United States exhibitors withdrew
from competition.   Professor Thurston, the chairman of the jurors
on agricultural implements, and consulting mechanical engineer to
the Exposition, stated that in design, finish, and smoothness of
operation the Canadian machinery was equal to anything in the
Exhibition.

Canada obtained 65 awards.   The list included seven
Horticulture.   different awards for Canadian grapes—a valuable tribute
to the climate of the country.   The small number of
awards is due to the fact that awards were only given to collective
exhibits, and not to individual exhibitors, or for each variety of fruit
shown.   With regard to the vegetable display, it was admitted that the
Canadian exhibit was greatly superior to any other.   Both fruit and

vegetables won the highest praise from the jurors for variety, excellence, and quality; and in this connection the report of the British Consul is especially interesting.

Live Stock. Canada more than sustained at Chicago her splendid record at Philadelphia in 1876 in this department, the live stock and poultry exhibited having secured more than one-half of the total prizes offered. In cattle, with 184 entries, Canada took 104 prizes, 17 medals, and 3 diplomas; against 532 entries of the United States, and 306 prizes and 13 medals. In horses, Canada had 96 entries, and 44 prizes, 2 gold medals, 10 medals, and 3 diplomas; the United States, 446 entries, 257 prizes, 6 gold medals, 12 medals, and 4 diplomas. In sheep, Canada, with 352 entries, secured 250 prizes, 5 silver cups, and 8 diplomas; against the United States' 478 entries and 193 prizes. In swine, Canada's 68 entries obtained 64 prizes, and the United States' 96 entries 67 prizes. In poultry and pet stock, Canada was awarded 501 prizes with 1,147 entries, and the United States 671 prizes with 2,453 entries. The grand totals were: Canada, 1,847 entries and 1,175 prizes; the United States, 4,005 entries and 1,494 prizes. This must be regarded as a very great success especially when the populations of the United States and Canada are taken into account. All the Canadian sheep and swine were bought by the Commissioner for Costa Rica.

Fish and Fisheries. The committee of jurors considered the Canadian fish exhibit the best and most complete in the Exposition. Twenty-four individual exhibitors also obtained medals.

Mines and Mining. No single exhibit in the mining building attracted more attention, and came in for more favourable comment, than the Canadian display; and the fact that there were 67 collective exhibits which took gold medals and diplomas in competition with other countries, speaks highly for the variety and richness of the mineral resources of the Dominion. The collections of ornamental and precious stones were much admired, and orders were subsequently received from two of the leading manufacturing jewellers of New York. The nickel ore exhibits were mentioned as being higher in grade than any other shown at the Fair. Canadian copper also obtained a flattering award; and the fine exhibit of anthracite and bituminous coal from all the mines in the North-West Territories, now being worked, attracted a great deal of interest.

Machinery. The machinery exhibit was a small one, but almost every exhibit took a prize, 43 gold medals and diplomas falling to the Dominion. Professor Thurston, chairman of the jurors, and consulting mechanical engineer to the Exposition, stated, in an address, that in design, finish, and smoothness of working the general machinery shown by Canada was equal to anything shown; and that, as compared with the exhibit made in 1876 at Philadelphia, Canada had made greater relative progress in manufacturing, since that time, than any other nation taking part in the Exhibition.

In this department Canada obtained 30 medals and
Transportation.   diplomas.   The Canadian Pacific Railway train was
referred to as the finest and most complete on exhibition, which reflects great credit on the position manufacturing
enterprise has reached in Canada.

The great development in the industries of the
Manufactures.   Dominion is illustrated very aptly by the following
return, taken from the census of 1891 :—

	1881.	1891.	Increase.	Per Cent.
Number of establishments ...	49,923	75,768	25,845	51·8
Capital invested	$165,302,623	$353,836,817	$188,534,194	114·0
Number of *employés*	254,935	367,865	112,930	44·43
Wages paid	$59,429,002	$99,762,441	$40,333,439	67·86
Cost of raw material	$179,918,593	$255,983,219	$76,064,626	42·3
Value of products	$309,676,068	$475,445,705	$165,769 637	53·5

Canada had a most interesting exhibit of manufactures, which secured
124 awards, and drew an appreciative statement from the president of
the jurors on textiles—a member of the Austrian Commission, and himself a manufacturer of high-grade cloths in Austria—to the effect that
the progress made by textile manufacturers in Canada had been more
rapid during the last five years than that of any other country show-
ing industrial products.   It will be remembered by many readers of
these pamphlets that Canada's display of manufactured articles at the
Colonial and Indian Exhibition in 1886 attracted much attention.

The educational system of the Dominion has a world-wide reputa-
tion, and the exhibits in that department were naturally, therefore, an
object of sustained interest throughout the course of the Exhibition.
191 awards were obtained by Canada.   No better evidence of the
excellence of the display can be had than that in a section supposed
to be the smallest among the Canadian exhibits, such a large number
of awards should have been secured.

PRAIRIE CHICKENS.

www.ingramcontent.com/pod-product-compliance
Lightning Source LLC
Chambersburg PA
CBHW022027190326
41519CB00010B/1621

* 9 7 8 3 3 3 7 3 2 8 6 6 5 *